Investigate

Sunlight

Sue Barraclough

www.raintreepublishers.co.uk
Visit our website to find out
more information about
Raintree books.

To order:

☎ Phone 0845 6044371

🖹 Fax +44 (0) 1865 312263

💻 Email myorders@raintreepublishers.co.uk

Customers from outside the UK please telephone +44 1865 312262

Raintree is an imprint of Capstone Global Library Limited,
a company incorporated in England and Wales having its
registered office at 7 Pilgrim Street, London, EC4V 6LB – Registered
company number: 6695582

Edited by Sarah Shannon, Catherine Clarke, and Laura Knowles
Designed by Joanna Hinton-Malivoire, Victoria Bevan,
    and Hart McLeod
Picture research by Liz Alexander
Production by Duncan Gilbert
Originated by Chroma Graphics (Overseas) Pte. Ltd
Printed and bound in China by Leo Paper Group

ISBN 978 0 431932 86 6 (hardback)
12 11 10 09 08
10 9 8 7 6 5 4 3 2 1

ISBN 978 1 406 24464 9 (paperback)
13 12
10 9 8 7 6 5 4 3 2

British Library Cataloguing in Publication Data
Barraclough, Sue
    Sunlight. - (Investigate)

A full catalogue record for this book is available from the
British Library.

Acknowledgements
©Alamy pp. **7** (David Cole), **8** (Martin Shields), **14** (Bruce Coleman
Inc.), **20** (Cultura), **27** (Iain Cooper); ©Corbis pp. **4** (Richard
Hamilton Smith), **5** (Craig Tuttle), **6** (Momatiuk – Eastcott), **11**,
25 (Tim Pannell), **12** (Don Hammond), **13** (Terry Eggers), **15** (ML
Sinibaldi), **23** (Ingolf Hatz/zefa), **24** (Robert Llewellyn), **26** (Chinch
Gryniewicz; Ecoscene), **28** (Manfred Mehlig/zefa), **29** (Bob Krist);
©Digital Vision p. **19**, ©Getty Images p. **18** (Robert Francis/Robert
Harding World Imagery); ©Istockphoto p. **16** (Paul Paladin);
©Pearson Education Ltd. p. **22** (Tudor Photography 2005).
Cover photograph of sun over clouds reproduced with permission
of ©Masterfile (Bruce Rowell).

Every effort has been made to contact copyright holders of
material reproduced in this book. Any omissions will be rectified in
subsequent printings if notice is given to the publishers.

Contents

Some words are shown in bold, **like this**. You can find out what they mean by looking in the glossary.

Sunlight

Sunlight is light and heat from the Sun. Sunlight shines on Earth and lights and warms our world.

All living things need sunlight to live and grow. If there was no sunlight, there would be no life on Earth.

Q What happens if plants do not get enough sunlight?

? CLUE

- What do plants need to grow?

 **A** If plants do not have enough sunlight, they do not grow well.

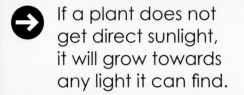

→ If a plant does not get direct sunlight, it will grow towards any light it can find.

Plants need sunlight to help them make food to grow. Plants take in sunlight through their leaves. They use sunlight with water and air to make food.

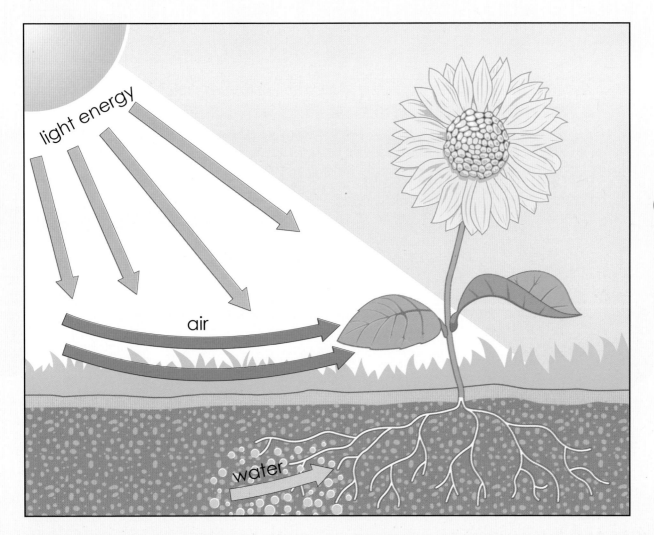

 Food is made in the plant's leaves.

Animals and sunlight

Animals need sunlight to stay warm. Many animals eat plants that use sunlight to grow.

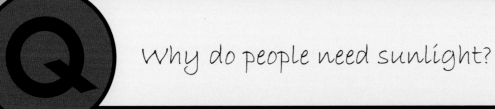

Q

Why do people need sunlight?

11

CLUES

- What would Earth be like without sunlight?
- Would it be light or dark?
- Would it be warm or cold?

A People need sunlight to light and warm Earth. Light and warmth are needed for plants to grow. People need to eat plants.

 Wheat is a plant that can be cut and used to make bread.

People grow fruit and vegetables to eat. Fruit and vegetables need sunlight to help them grow.

The Sun

The Sun is a **star**. The Sun is the closest star to Earth. It is a huge, hot ball of burning **gases**. The Sun throws out **beams** of heat and light. Never look directly at the sun. It can damage your eyes.

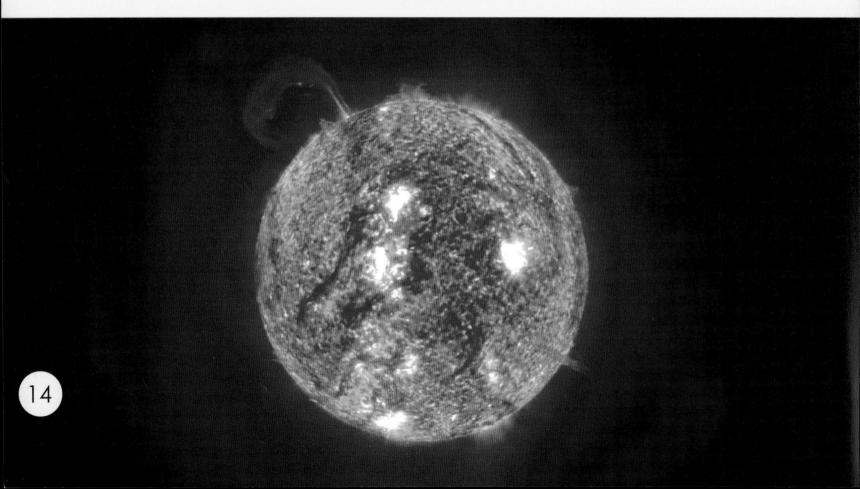

The Sun seems to rise into the sky in the east. This is called sunrise. The sky starts to get light at sunrise. Then the Sun seems to move across the sky, and sets in the west. This is called sunset. The sky grows dark after sunset.

 Does the Sun move around Earth?

15

A The Sun does not move around Earth. Earth moves around the Sun.

At night, Earth is blocking out the sunlight. This means that part of Earth is in **shadow** and it is dark.

16

As Earth turns, light from the Sun shines on different parts of Earth. Where sunlight shines on Earth, it is daytime.

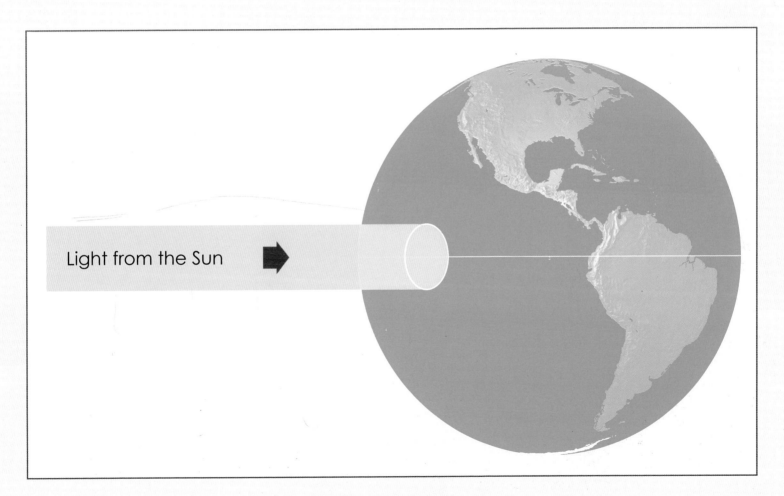

Light from the Sun

Hot and cold

When the Sun shines on Earth, it is warm. The middle of Earth is nearest to the Sun so it is hotter. The weather is always hot in the middle of Earth.

At the top and the bottom of Earth are places called **poles**. The poles are furthest away from the Sun. The weather is always cold at the poles.

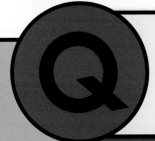

Does the weather stay the same in all parts of the world?

Seasons

No. In some parts of the world there are four different seasons. Each **season** has different weather.

The seasons are spring, summer, autumn, and winter.
The Sun causes the different seasons.

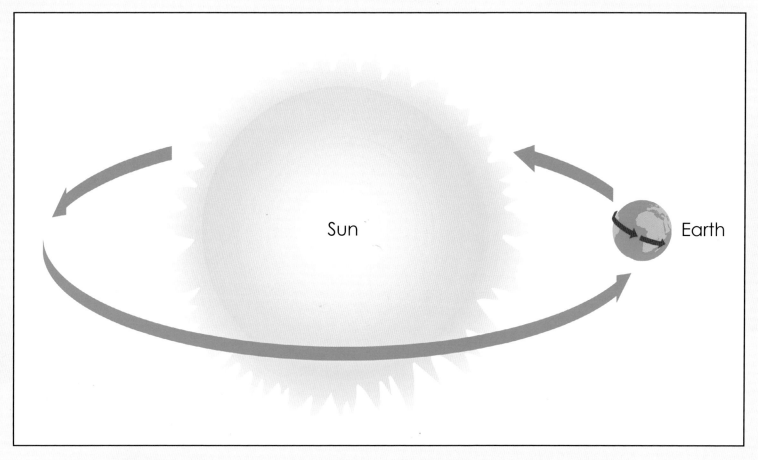

 This journey takes one year. When Earth is tilted away from the Sun, the weather is colder and it is winter.

In spring, the weather gets warmer. The days get a little longer. There are new leaves on the trees. Flowers start to grow. Many birds build nests to lay eggs.

Summer is the warmest season. The days are long and light. It does not get dark until late in the evening.

Q

Which seasons have cooler weather?

Autumn and winter.

In autumn the weather gets cooler. The days get
shorter. Leaves on many trees turn from green to
brown. Squirrels collect nuts to store for winter food.

Winter is the coldest season. The weather can be snowy and icy. Plants stop growing and many trees have no leaves. The days are short and it gets dark early in the evening.

The power of the Sun

The Sun gives Earth heat and light. The heat of the Sun can be collected by solar panels. Solar means "from the Sun". Solar panels can be used to heat water.

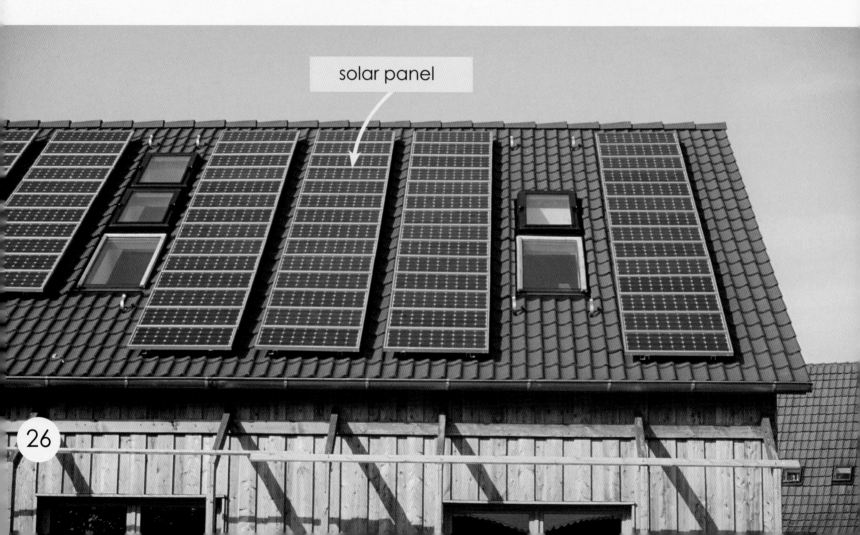

solar panel

The power of the Sun can also be used to make **electricity**. Electricity makes things such as kettles and televisions work.

The solar panels on this road sign help to light up warnings to drivers.

Even though the Sun is far away, it makes life on Earth possible. Sunlight means that plants grow. The different seasons and weather, such as rain, snow, and wind, are caused by the heat of the Sun.

Sunlight brings light, warmth, and colour to our world. Without sunlight, there would be no life on Earth.

Checklist

Living things such as plants, animals, and people need sunlight.

Sun facts:

➠ It takes one **year** (or 365 days) for Earth to travel around the Sun.

➠ It takes one **day** (or 24 hours) for Earth to turn.

➠ The Sun is 150 million kilometres (93 million miles) from Earth. It is so far away that it takes 8 minutes for sunlight to travel to Earth.

Glossary

beams lines of light that shine from a bright object

day length of time that is 24 hours long, usually from midnight one night to midnight the next

electricity form of energy that makes light and heat and can make machines work

gas air-like substance that is usually invisible. Gases are not liquids or solids.

poles parts of Earth that are furthest north or south

season time in the year when a type of weather usually happens

shadow dark area that is made when something blocks out the light. When sunlight hits something it cannot shine through, it makes a shadow.

star large ball of burning gas in space. Most stars are seen as points of light in the night sky. The Sun is the only star we can see during the day.

year length of time that is 12 months (365 days) long

Index

Plants and sunlight

Plants need sunlight to grow. They use sunlight to make food. The heat of the Sun helps flowers grow and open. Sunlight helps fruit to ripen.

Sunflowers grow towards the Sun.